THE
BIODYNAMIC
FARM

THE BIODYNAMIC FARM

Developing a Holistic Organism

Karl-Ernst Osthaus

Floris
Books

Translated by Beate Buchinger

First published in 2004 in German under the title
Weg aus der Krise by Pelagius Seminar, Bad Liebenzell

First published in English in 2010 by Floris Books, Edinburgh
Second printing 2016

British Library CIP Data available
ISBN 978-086315-766-0
Printed in Poland

Contents

Acknowledgments

This book finally came into existence after much imploring by my Russian, Polish, French, Austrian and German friends. However, without the continuing encouragement of Anne-Marie Rösner, I would not have seen the necessity of writing down my knowledge and perception acquired in a lifetime of farming. For this and for her assistance in my writing, I thank her ever so much. I also thank all my friends, who have helped me in various ways to complete this project.

Karl-Ernst Osthaus
Unterlengenhardt 2004

Foreword

Biodynamic agriculture is an approach to the earth that not only incorporates the ethos of organics but an all- embracing attempt to understand the widest and most intimate connections of nature. Every plant and animal, each habitat, the weather, the sunlight, the moon and all the stars in the sky, form part of a great living universe. Nor is any portion of it needlessly there. During the course of lectures which Rudolf Steiner gave at a conference for farmers in 1924 and which has become a continuing source of inspiration for the worldwide biodynamic movement, he continuously sought to extend the imagination of his listeners. 'We shall never understand plant life,' he said, 'unless we bear in mind that everything which happens on the earth is but a reflection of what is taking place in the cosmos.'[1]

Providing a wheat plant with sufficient nitrogen alone will not suffice. Only when its roots can thrive within a living soil, its leaves and stem experience life in the wind and rain and its ears ripen in the warmth and light of the sun can a wholesome plant develop. The skill of the farmer is not primarily to feed but to awaken the sensitivity of plants to the subtle influences that continuously stream in from the stars. In his course of lectures Rudolf Steiner gave clear and practical indications as to how this might be achieved. The biodynamic preparations described in detail later in this book serve to stimulate the processes whereby the physical substances of the earth can be enlivened and through which the plants can begin to 'hear' what streams from the stars.

The farm itself, instead of being simply viewed as a means for producing so much grain, milk or meat should really be considered as a living organism held within its geographical confines but holding the boundless universe in its arms. Rudolf Steiner said:

> A farm is true to its essential nature, in the best sense of the word, if it is conceived of as a kind of individual entity in itself — a self-contained individuality. ... Whatever you need for agricultural production, you should try to possess it within the farm itself (including in the 'farm,' needless to say, the due amount of cattle). Properly speaking, any manures or the like which you bring into the farm from outside should be regarded rather as a remedy for a sick farm. That is the ideal. A thoroughly healthy farm should be able to produce within itself all that it needs.[2]

This core principle of biodynamic agriculture lies at the heart of what the author of this book is wanting to expand upon and share with his readers.

It has been a common assumption among biodynamic and organic farmers that in order to provide sufficient fertility a farm needs to ensure a livestock population of about one cattle unit per hectare and to have a diversity of livestock. This has proved a very successful principle to work with even though today the ratio of 1:1 is considered to be on the high side for arable farms. For the author, however, the number of livestock units per hectare is not the most significant thing. Instead he focusses on their relative numbers in a more universal sense. He recommends, for instance, that 12 cows, 4 horses, 6 pigs, 10 sheep and 120 hens is the correct number and a combination for a small farm of about 60 hectares (150 acres). The reason for this particular proportionality is drawn from an understanding of the cosmic significance of these numbers, the nature of the animals in question and his lifelong experience as a farmer.

This deeply intuitive approach to the way the farm animals, as well as wild animals, plants and landscape features, contribute to the spirit of the farm is a remarkable quality, and one which many biodynamic farmers aspire to. His is also a deeply meditative approach and it is this quality which Rudolf Steiner sought to convey when speaking as he did about the spiritual in nature.

Rudolf Steiner was a philosopher but he was also a scientist. He sought to show that just as we can objectively investigate the nature of rocks, plants or natural laws, we can also explore the spiritual forces and beings which stand behind them, if we have the organs to perceive them. In his book, Knowledge of Higher Worlds, Steiner shows how through rigorous inner training these new capacities can be developed. By following this path himself he was then able to share insights into the spiritual world based on his own objective observations. The results of his investigations are then presented through the many thousands of lectures which he gave on all aspects of human existence including agriculture, education, medicine and much else besides. This new spiritual science was for Steiner not an alternative but a complement to natural science. The two belong together as two sides of the same coin.

In seeking to bring across his observations of the spiritual world, new words, often imported from the theosophical tradition, had to be used to describe particular phenomena. Two such words are 'etheric' and 'astral.' Etheric refers to the living vitality that pervades all plants and animals, while 'astral' is connected to the inner drives of hunger and thirst and with the world of animals. A closer look at the kingdoms of nature will reveal something of these differences. Minerals are lifeless, plants are alive and vital, animals are alive and vital but can also experience hunger, thirst and pain. Thus the mineral world is purely physical, plants are physical and alive (etheric), animals are physical, alive and conscious (astral). Human beings are physical, alive, conscious and in addition are aware of themselves (spirit). An in-depth explanation of this can be found in Theosophy, another basic book by Steiner.

Anthroposophy (the name given by Steiner for the whole field of spiritual scientific research) underlies everything which Osthaus describes in this book. Since he was a farmer, however, he writes out of a direct and immediate experience of the effects of the spiritual world working through nature and his farm. Wise farm management consists of allowing nature to express herself within a reverential yet firm direction given by the farmer. The greatest natural diversity is allowed to flourish on the farm while at the same time serving the purpose of producing a rich surplus of food and other products for human beings.

By working intimately with nature and the spiritual world and employing the special techniques of biodynamic agriculture, food can once again be produced that provides true nourishment, a landscape can be formed that heals and a context can be created within which human beings can grow in peace. This is the attitude with which Karl-Ernst Osthaus addresses many different issues — why cows have horns, how manure should be handled and why woodland areas are so essential to every farm.

Bernard Jarman
May 2010

Introduction

The natural sciences have brought us the clarity of mind and an ability to think clearly, as well as valuable technical acquisitions. However, through their one-sidedness they have blocked our ability to view the spiritual world. In order to salvage living nature — including man — they need to be supplemented by spiritual science, for only spiritual science can render the knowledge the natural world needs in order to continue to live in a healthy manner. Of all the sciences, physics is the one that leads us away from the dead-end road of materialism into which we have slithered, and back into the world of the spirit, the foundation of all creation.

Professor Dr Max Plank (1858–1945), winner of the Nobel Prize, expressed the following thoughts in a lecture held in Florence, Italy, in 1944, concerning the nature of matter. He was one of the outstanding physicists of the early twentieth century. He is particularly significant among German physicists, primarily due to his upright character and his imperturbable, clear-cut manner.

Gentlemen! As a physicist I have served all my life the most sober, matter-of-fact science, being the exploration of matter. Hence, I am sure, not to be considered a dreamer. And this is what I have to say after my research on the atom: there is no matter as such! All matter comes into being and exists by means of a power, which brings and keeps the parts of the atoms in resonance, thus forming the tiniest solar systems. However, since there is neither an intelligent nor an eternal force within the whole universe, we must assume an intelligent being of spiritual nature behind this force. This spirit is the absolute base of all matter! Not the visible and transitory physical matter is what is real, is true, rather it is the invisible and immortal spirit that is reality! But since spirit as such alone, likewise cannot exist, since every spirit belongs to a being, we are compelled to assume spiritual beings. However, spiritual beings cannot come into existence by themselves alone, rather they have to be created. This mysterious creator I do not feel embarrassed to call God, as all cultures of the earth in earlier millenniums have called him. So you see how in our days, in which nobody believes in the spiritual as being the ultimate base of everything created, hence living in bitter estrangement to God, it is the tiniest and the invisible which resurrects the truth from its grave, the grave created by materialistic illusion. This will change the world. Thus the atom has opened the door for humanity to the lost and forgotten world of the spiritual.

Paracelsus also broadens our view. He writes:

From this it follows, that the human being is the smallest world. That is, he is the microcosm due to the fact that he is the whole world since he is an extract of all stars, all planets, of the whole firmament, of the earth and all elements.

These two quotations — one by a modern physicist known throughout the world, the other from the Middle Ages by Paracelsus — should promote a questioning attitude. Both direct us towards another world, which is at least equally real as the world in which we live presently. This intuitive feeling should form the basis for planning our actions in the future. In order to heal the earth, farmers of the future have to be in accord with this other world, so that the development of mankind, which occurs in the realm of the spiritual and the soul, is possible.

Since the developmental path of mankind is of a flowing nature and everything changes, old knowledge has been forgotten — especially so under the influence of science. For that reason, the aim of this book is to help find the future way of farming, while retaining old knowledge. However, a new way of farming does not imply a return to former agricultural methods. What must be termed 'new' is the broadened way of thought, which encloses the whole universe with its hierarchies, and which can and should be intensified to spiritual perception.

The Wider Background

Can mankind regulate its affairs so that its chief
possession — the fertility of the soil — is preserved?
On the answer to this question the future of
civilization depends.
Sir Albert Howard, An Agricultural Testament, 1940

This clarion call from a renowned pioneer of organic agriculture some seventy years ago is even more urgent in the catastrophic situation of today. In 1924, Rudolf Steiner, the founder of anthroposophy, expressed similar views during his course of lectures on agriculture, with the following words:

Even materialistic farmers nowadays ... can calculate
in approximately how many decades their products
will have degenerated to such an extent that they
can no longer serve as human nourishment. It will
certainly be within this century. This is a cosmic issue
as well as an earthly issue. Precisely from the example
of agriculture, we can see how necessary it is to derive
forces from the spirit, forces that are as yet quite
unknown. This is necessary not only for the sake of
somehow improving agriculture, but so that human
life on Earth can continue at all, since as physical
beings we depend on what the Earth provides.[3]

And furthermore:

Humanity has only two choices: either to start once again, in every field of endeavor, to learn from the whole of nature, from the relationships within the whole cosmos, or to allow both nature and human life to degenerate and die off. There is no other choice.[4]

We can experience how today, some eighty years later, Rudolf Steiner's words are fast becoming reality. It is also worth including at this point a comment made by Rudolf Steiner in response to a question by Ehrenfried Pfeiffer: 'Why is it, that in spite of theoretical understanding, the will to take action out of spiritual insight is so weak?' Rudolf Steiner gave the surprising and remarkable answer:

This is a nutritional problem. Food as it is today can no longer provide human beings with the forces necessary for manifesting the spiritual within the physical. It has become impossible to bridge the gap between thinking and action. The plants serving as food no longer contain the forces people need.[5]

In pondering these words we can gain an inkling as to the extent to which humanity has declined, not only physically but also as regards its spiritual capacities. All manner of illnesses are appearing and we are being forced to acknowledge them as being part of normal life. What is offered as food today is not nourishment in the true sense since it lacks vitality. This is demonstrated by the low level of minerals, trace elements and vitamins it contains. Many people are dependent on nutritional supplements. A healthy state of consciousness, however, is only possible if plants can transmit their vitality directly to human beings. Plants only develop such vitality when grown in healthy soil. During the course of the last hundred years these healthy soils have been almost totally destroyed. If human life and culture is to have a future we have to start with the soil. We need to hearken once again to the word of Hippocrates: 'Our food should be our medicine. Our medicine should be our food.'

There is repeated reference to spiritual (invisible, non-physical) forces in the following text. According to Max Planck: 'Only by accepting the existence of spiritual forces will we be able to gain a full understanding of the kingdoms of nature and begin to work with them.' What kind of forces do we then need in order to produce food that can provide a healthy foundation for mankind's continued existence?

The forces are of a spiritual nature, and it was Rudolf Steiner's great contribution to the twentieth century to describe these spiritual forces in a way that we can understand today. He described the different kingdoms of nature in the following way.

1. Mineral kingdom consisting of dead matter obeying physical laws.
2. Plant kingdom consisting of physical substances imbued with life. These life or etheric forces structure the matter in plants and determine their pattern of growth.
3. Animal kingdom consisting of living physical substances and permeated by astral forces. Astral forces are soul forces of diverse kinds. The soul of an animal allows it to have sensations and to act out of these sensations. Additionally, each species of animals has a collective group soul that guides each animal species. This is partly expressed as the instinct of the animals.
4. The human kingdom incorporates all the qualities of the above mentioned realms as well as a further spiritual element — the human 'I.' It is through this spiritual element that human beings are conscious of themselves and can speak of themselves as 'I.' The human 'I' has some resonance with the animal group soul which guides the species from the spiritual world. A part of this 'I' is the higher self, which Friedrich Schiller called our 'divine spark.' It is our conscience, and it enables

us to practise self-control and to continue developing. Through this higher self we are free of instincts and are able to determine our actions freely.

This ability is at the same time a responsibility: we are invited to conduct our lives to the best of our ability and conscience. At the end of Part 2 of Goethe's Faust, it says: 'Whoever fervently strives, can be redeemed.'

The goal of creation is the free human being who no longer simply follows the Ten Commandments as was necessary during an earlier cultural period. Our individual path of development leads us towards achieving self-determination and moving from 'Thou shalt' to 'I will.' Our goal is to serve the divine order out of our own free will, and by so doing find individual fulfilment. People have appeared throughout history who were able to lead mankind to higher forms of culture by their example. They are often called saints. These are our precursors and ideals.

All this sets the wider background to the task of the farmer today. This book describes how to work in harmony with nature and provide the basis for the continuing development of the earth, of plants, animals and the human being.

Soil and consciousness

Moral decline, depression, aggression, lack of will power, weakness, fatigue — these phenomena appear in our modern industrial society. Parallel to this is the loss of humus and soil degeneration throughout the world. There may well be a connection between the two. Recent research has shown that there is a connection between the health of soil on the one hand and the health of plants, animals and human beings on the other — right through to changes in consciousness.

There is a remarkable correspondence between the pH-value of 7–7.2 in healthy human blood and that of healthy soil. Today's agricultural methods, however, tend to make soils more acid. This raises the question as to what happens when our food is derived primarily from acidic (that is degenerated) soil? Human health is highly influenced by the quality of certain protein molecules, the DNA, which also direct the immune functions of cells. Cell protein is the substance which gives life, regulates growth and keeps our cells healthy. Protein and protoplasm are the bearers of life processes.

Due to the damaged state of today's soil — which is primarily the result of chemicals employed in modern agriculture — the quality of protein is no longer what it should be. That means the DNA in the body is being damaged by debased food, with its reduced and hence inadequate effect on the organism. The many kinds of cell toxins which find their way into the human body damage not only ourselves, but also the plants and animals upon which we depend for our nutrition. This results in an increasing number of metabolic and deficiency diseases as well as a weakened immune system.

It is striking that certain lactic acid forming bacteria are found around plant roots, which are identical to those found in the human digestive system. There is a certain conformity between soil metabolism and the human digestive system. If soil life disappears, or if the soil becomes so degenerated that it can only serve as an artificial substrate for unnaturally bred plants, it will of course have adverse consequences for the health of the plants growing in such soil, to the point of showing severe deficiencies. Such plants used as food will likewise affect the physical and mental well-being of humans and animals. Our state of health or illness is largely determined by our nourishment and surroundings. The increase of mycologically (fungus) caused diseases may well be a result of the deteriorating quality of our food.

New research shows that each cell radiates an extremely weak light.[6] This radiation makes it possible to examine the quality of proteins. The light creates order within cell systems. These are cosmic systems of order, in harmony with nature, which are the foundation of living soil. But they are also active in plants, animals and humankind. What happens if these protein structures are damaged by human interference? The light signals change, and the whole cosmically ordained system of order changes. Tremendous changes have taken place within soil through human intervention, and it is no longer possible to speak of the soil as being a system with its proper, natural organization. This disorder manifests in plants and their growth. For instance, the silicon content of plants has dropped by more than 40% in the last fifty years. So a plant of today cannot be compared with the same kind of plant of fifty years ago. Other changes compound this: for instance, the degenerated proteins are passed through the food chain into animals and human beings. There are also the destructive effects of electromagnetic radiation, which have increased exponentially in the last few years.

One can say in a simplified way that proteins are structured by light. Enzymes containing phosphorus have an important task in this process. In the body, proteins are also destroyed by these enzymes, thus setting the light free again. This light within the body is the energy, which life needs. This light energy of life is carried into the nervous system, into the substance of the brain, where it serves our consciousness. It is not for nothing that we call a gifted person 'bright.'

Degenerate soil, mineral and nitrogen fertilizers, the breeding of plants for quantity rather than quality, result in proteins of inferior quality whose amino acids are no longer balanced. This leads to a lack of light in our digestion. This lack even has its effects on our mind. Many abnormalities of emotions or of consciousness are the result of this.[7] The dominance of materialistic intellect, devoid of all spiritual content, is perhaps also due to this diminished light metabolism.

Modern agriculture adds another element of destruction. The present way of tilling the soil and adding manure leads to putrefaction on a huge scale. Putrefaction occurs when protein decomposes in anaerobic conditions, that is, when lacking oxygen. Hydrogen sulphide, which is a strong nerve toxin, develops. There is a direct connection between the increase of putrid decay (starting in the rumen of the cow as a result of feeding silage and concentrated feed), and the increase in psychiatric illnesses. The French ecologist, André Voisin, pointed to this connection.[8] The negative effect on mental health — depression, aggression and many other serious conditions — caused by soil destruction and slow poisoning are obvious.

Furthermore, for over a century, decreasing amounts of silicic acid have been found in plants. Between 1880 and 1930 the decrease was 30%. From 1950 onwards, the decrease accelerated due to the more intensive use of mineral fertilizers, chemicals, the introduction of liquid fertilizers and the increase of electromagnetic radiation. These factors continually inhibited plants' intake of silicic acid from the atmosphere. Consequently there is a marked reduction in plants' intake of light, as this process requires silicic acid. Protein production in plants is dependent on light intake and its effects. So, as light metabolism is reduced in plants, the quality of its proteins deteriorates.

As long ago as 1952, biochemical research at the University of Chicago showed that healthy forms of protein in plants could no longer be found anywhere in the world. This has further consequences. Because of light deficiency and plants' consequent inability to produce wholesome protein, they also contain fewer minerals, trace elements, enzymes and vitamins.

Today, protein degeneration has advanced to such a degree that it has also affected animals and human beings. In cows with BSE, sheep with scrapie and human beings suffering from Alzheimer's, MS, Parkinson's disease and Creutzfeldt-Jakob disease, degenerate forms of protein were found in the nervous system and brain. These diseases, previously almost unknown, have only become widespread in the last fifty years. They are likely to increase in the future, unless our way of farming stimulates the intake of silicic acid in plants,. This is possible through the application of biodynamic herbal preparations, especially the dandelion preparation (see p. 80). They help to increase the silicic acid content of plants, which is visible through their greater luminosity.

It is clear that the way we farm and treat our soil is of fundamental importance for the continued existence of humanity. In this book we describe proven ways of healing the soil, which are dependent on smaller, human-scale diverse farms working in harmony with nature. Only on such farms is it possible to live with and observe nature with sufficient intensity.

Perspectives for farming

The development described in the previous chapter can lead us to question the whole basis of present-day scientific views. For if these views continue to be applied to the earth, to nature and to human beings, we can predict a decline which will ultimately lead to the destruction of our world.

Throughout history, people's views were based on divine-spiritual beliefs which directly influenced life on earth in its practical and social aspects. These old civilizations always declined and degenerated when their ties to the spiritual world loosened.

In this light, the necessity of gaining an insight into the powers of the non-physical, divine-spiritual world becomes clear. Our understanding of nature must be widened to an all-embracing knowledge of the world, including its non-material aspects. Just as any one organ within a living organism exerts influence upon the others, so we must try to see nature and the earth as a living organism which reflects the laws of the cosmos.

This begins with the creation of agricultural organisms that contain all realms of nature — soil, plants, animals as well as human beings. For a relationship to the cosmic-spiritual powers that allow new forces and processes to work can only be made by living organisms. In an agricultural organism the individual parts will influence each other in a perfectly natural way.

Such an agricultural organism permits cosmic laws to work and permeate the earth, forming a kind of microcosm, just as the human being and its organs form a microcosm.

Animals on the farm

As the heart is the central organ in the human being, so cows can be regarded as the central organ within the agricultural organism. To reflect cosmic laws of number, ideally there should be twelve cows or, if the farm is very large, a multiple of twelve, corresponding to the twelve signs of the zodiac. The number of other animals is then determined by their tasks and needs. Thus a farm with twelve cows and their offspring needs about 30 hectares (75 acres) farmland and 30 hectares grazing land. To farm this, four horses (two teams) and their offspring are needed. Pigs are necessary as waste utilizers. They are important for cleanliness and hygiene on the farm, preventing decay and the development of hydrogen sulphide. Four to six pigs are needed for twelve cows on 30 hectares farmland and 30 hectares grazing land. The number of sheep will depend upon the soil, the climate and the needs of the people (clothing, etc.), but about ten sheep are suitable for a 60–70 hectare farm. Then fowl are needed. An old rule is two chickens per hectare (about one per acre). At first sight this doesn't seem many. However, when trying to feed this number of chickens on by-products throughout the year, it becomes clear how difficult it is to feed even this small number of chickens well. Then there are geese, who live in symbiosis with the sheep and can graze with them. They serve as pest control on the pastures and supply feathers for bedding; their number is determined by need. Similarly the number of ducks is determined by their service as pest control on pastures, meadows and swampy areas. Also a few guinea fowl on the farm will keep rats down, since rats cannot stand their screeching. A few cats are needed to keep mice down, and a good guard dog to note any irregularities on the farm. The dog is a kind of extended sensory organ of the farm family.

Hedges

But there are other creatures of importance as well: primarily birds, bees and ants. In order to have sufficient numbers of these, the farm must be arranged organically by planting hedges and trees throughout. Every meadow, field and pasture should be surrounded by hedges of various trees, shrubs and bushes, which bear flowers and fruits; hawthorn is especially beneficial. Every tree and shrub has its particular characteristic value. The hazelnut is especially important in protecting against unhealthy radiation of all kinds, and serves as additional feed for cattle. Hawthorn is important for birds, who can build their nests in it, well protected from birds of prey and cats. In spring the smell of hawthorn flowers is a stimulus to other plants. In autumn and winter the red berries serve as food for birds and can also be used medicinally. Sloe, with its runners, creates a dense and very thorny hedge, giving protection to birds. In early spring its flowers develop a lovely scent, stimulating plant growth, and in winter its fruits serve as feed. Elixirs and syrups can be made from the fruits. The wild rose is especially rich in scent and its fruits can also be used for elixirs and jams.

The rowan tree should also be included as a hedge plant. It has abundant fragrant flowers and fruit. Fruit from the grafted rowan tree can also be made into jam. Various kinds of willow can be planted because the leaves are especially important for cows and the pollen is essential for bees.

The bird cherry, a small tree or shrub, produces fragrance

from its white clusters of flowers in spring and bird feed from its black berries in autumn. The fruits can also be used for making juice. The bird cherry is prolific — its vigorous growth and dense foliage tends to suffocate neighbouring plants, so that within a few years you will have only bird cherry in your hedge, which should be prevented.

The alder tree is important as well — both red alder and white alder — as a collector of nitrogen. In its early life the alder grows faster than all other trees, aiding the other trees as nurse-plant. Alder foliage is a good substratum for the large stinging nettle (Urtica dioica) of which there cannot be enough. Nettles regulate the iron content of the soil. They are necessary for biodynamic preparations (see p. 75) and can be cooked in a similar way to spinach, as well as providing food for pigs and for young chicks.

It is good to integrate lime (linden) trees into the hedge because they develop an intense fragrance and attract lots of bees. Oak should also be included because it exerts a strong influence on ground water with its taproot and its foliage offers a habitat to many insects. The ash tree, the tree of the sun, is also an important hedge plant. Its foliage can be used for making hay; because of the assimilated sun influence in its foliage, this hay is an excellent health booster for farm animals.

The birch, the tree of light, should also be present in the hedge. Generally it does not need planting as it is prolific. As it has flat roots and a great need for water, there should not be too many in the hedge. In spring it is the first tree to turn green, and with its spicy fragrance encourages the growth of other plants. Its pollen in early spring is vital for bees.

The common hornbeam is also a good hedge plant. Only loosing its leaves in the following spring, it offers protection to many animals in winter. Common dogwood should also be planted for beauty's sake and for fragrance and fruits.

When preparing to plant a hedge, the ground should first

be ploughed as deeply as possible, piling up the soil to make a kind of wall. The higher it is, the better. A greater vitality develops in this wall, giving the hedge better conditions for growth. The higher this wall is, the less the roots of the hedge plants will spread into the adjoining fields. If possible, a furrow can be ploughed on each side and the soil put on the wall before planting. This will further prevent the roots from growing into the adjoining fields. Foliage will gather in the furrow in autumn, rotting and soaking up water, stimulating additional growth in the hedge. Earthworms and other small creatures will be encouraged to settle there.

In areas rich in wildlife, the hedge will need to be protected by a wire-netting fence for about five years. When caring for the hedge, the alders have to be cut back first because they are the fastest growers and will otherwise suffocate the other plants. After being cut, the alders will react with a strong growth from their stumps, making the hedge dense at its base.

It will be necessary to cut the hedge right back every seven years. As the branches then sprout from the base, the hedge becomes dense. A few trees can be spared: oak, ash or lime as needed. A lot of wood will be produced in doing this, easily covering any need for firewood. Hedges also supply foliage hay for the herd. Branches should be pruned around St John's Day (June 24th), gathered into bundles and hung from the roof to dry. This foliage is excellent winter feed, of which there cannot be enough.

The composition of the hedge will of course vary according to the nature of the soil and the region. Hedges regulate the groundwater system and can even prevent a lowering of the groundwater level. The scent of the flowering hedge stimulates the growth of all other plants. Crops, pastures, meadows and wooded belts will mutually benefit from this harmony.

The distance between hedges should be primarily determined by the radius of activity of small birds, which is about 50 metres (165 feet), giving a maximum distance of 100 metres (330 feet) between hedges. The greater the diversity of the hedge shrubbery, the greater the diversity of birds living there. By planting walnuts, chestnuts and lime (linden) trees at various points, the whole landscape can be enhanced, which is easily observed. Birds, having a suitable habitat, will come in great numbers and diversity, and accomplish the tasks nature set them.

Hedges have another very important function. They act as windbreaks, calming the air and thus capturing carbon dioxide which is directly above the surface of the soil. This is important, as it stimulates plant growth.

It is also a good idea to plant copses of coniferous trees. Birds have a special relationship to these trees and modify their behaviour, which can prevent crop damage by feeding birds.

Insect life

Bees and ants are vital for the sound and healthy growth of plants. Bees have a 'bee poison,' a refined kind of formic acid. Formic acid is a bearer of life or etheric forces. In gathering nectar, as well as bringing their 'poison' to every plant, the bees bring etheric forces. They also produce and spread so-called epiphyllic yeast on the plants. When cows eat these plants, this yeast enters the first stomach and assists digestion.

Ants permeate trees and soil in woods with formic acid. This gives the soil, the humus of the woods, its specific character. Ants take care of aphids in trees. The liquid the aphid produces is a vital food for the ant. So they even bring aphids into their anthills, keeping them there like cows, 'milking' them daily. In this way the number of aphids is regulated.

Trees, aphids and ants form a symbiotic community, each helping the other. In many parts of our countryside ants have disappeared because of the influence of electromagnetic radiation (mobile-phone masts, television, radio, overland power lines, etc.). If there are no ants, humus will not be produced in forests, leading to decay and trees dying.

The care of birds and insects is really of fundamental importance to an agricultural organism. Only through the harmonious interaction of all realms of nature — the micro-organisms in the soil creating carbon dioxide, the fragrance of flowers and herbs, the presence of a large and diversified insect population, and abundant birdsong — can a healthy and resistant basis be created for all plant life, and thus also for the health of animals and humans.

Farm Animals

Constructing suitable barns

The needs of the animals as well as the relationship between man and animal have to be considered in the construction of barns and stables. Since ancient times animals have accompanied human beings in their development and evolution, continually being drawn closer and growing tamer. This has to be considered when keeping animals, and is especially important with cattle. Cows represent the digestive system of the agricultural organism. They need to be kept in optimal conditions for their digestive systems. These conditions are warmth, quiet, dim light and fresh air without draughts. In an ideal barn the cows are tied to their manger and supplied with plenty of rye straw as bedding. The ceiling slants slightly downward, dropping from the outside wall to the interior, where the cows heads are. It is 2–3°C (3–5°F) warmer in the cows area than in the rest of the building. The warm breath of the cows rises, moving through the ceiling into the hayloft, which has an opening for ventilation at each end of the barn. Thus the exhausted air rises, passes over the hay and leaves the building without any draught. This also avoids the cows breathing each other's expired breath, which they do not like. Half-high boards can be placed between cows, grouping them two by two. This will further calm the airflow and give the cows a strong feeling of security.

These conditions for keeping cows are the best for a good relationship between man and animal. Open stalls for cows, which are a legal requirement in some countries today, do not foster domestication of the animals, but encourages their wild nature. It is because of this increasing wildness that cattle are dehorned. This in turn influences the quality as well as the quantity of milk.

In the old farmhouses of Lower Saxony the horses were kept at the end of the building in loose stalls. Horses are built for action: they represent the limbs of the agricultural organism. Hence their stables allowed freedom of movement, giving them plenty of air and light. The whole farmhouse was an image of the proportions of the human body: at one end lived the head of the farm, the farmer with his family and farmhands; in the middle part lived the cows, representing the digestive system; and at the far end lived the horses, the animals of the limb system. This farmhouse construction grew out of wise, ancient religious traditions. Pigs and sheep were kept in nearby separate buildings in loose stalls. The proportions of the main house reflected the golden section. Such a building was always in harmony with the landscape, enhancing the countryside, and never seemed to be an alien element.

Cows within the agricultural organism

Within an agricultural organism the cow is particularly important. People have kept cattle since ancient times. An ancient Persian legend recounts that in the beginning of the world the archetypal human being and the archetypal cow were in the spiritual world. They conferred about the world to be created and about man who was to come into being. Finally the cow said, 'In order that you, man, can evolve, I will help you and sacrifice myself for you.'

With this legend as a background it is easier to understand how cattle has accompanied humanity in its development from the beginning, always sacrificing itself, especially nowadays. The cow is the archetypal animal of sacrifice.

Cow and farm must be a unit. That is, the cow herself should be born and raised on the farm and be of local stock. She is the fundamental element of a farm: soil, farm and cow forming a union. She should only be fed what grows on the farm: grass, hay, beets, kale, foliage, foliage hay and even straw, since her digestive system can utilize rough feed.

During the cow's complex digestive process, the contents of the four stomachs and intestine are permeated with etheric forces as well as astral and spiritual forces which come from the cosmos. After 36 hours, the enhanced intestinal content is excreted and becomes a precious fertilizer of semi-solid consistency, with a delicate, not unpleasant smell. The forces contained in this fertilizer enter the soil, the plants, and through them nourish the human being. In this way even barren soils can be brought to fertility. The cow's complex digestive system means that its dung will bring more vitality to the soil than that of any other animal. From this viewpoint we can understand why, in ancient religions, the cow was regarded as a sacred animal; this tradition survives to this day solely in India. We can only begin to understand how nature and humanity can live healthily when this most important capacity of the cow's digestive system is once again valued and applied. This understanding will in turn lead to discovering how best to keep cows so that they enhance the life forces of the soil and promote the right development of humankind.

Anthroposophy tells us that the whole animal realm evolved out of the archetypal, spiritual human being. From this viewpoint, each animal has developed in some specialized way as a kind of sacrifice, to enable human beings to grow in our unique way with consciousness of self and the world.

These ideas may help to foster a better attitude towards animals, to see them almost as brother or sister, and perhaps even to feel that it is our duty to include animals in our developmental path — to draw them to us, to tame them so that ultimately, when the earth has fulfilled its task and human evolution has reached its goal, the totality of life can be united in a higher spiritual realm.

Nowadays farm animals are pushed away from us, treated as objects, as means of production. A new way of animal husbandry in accord with nature has to be found. A more intimate observation of nature shows that all animals which live close to people are basically oriented towards us, waiting, as it were, to see whether humankind will ever really understand them and provide them with the best life conditions. Birds demonstrate this most clearly.

How can the breeding of cattle be best managed? A view based only on science is no longer sufficient. We have to consider the more subtle parts of the human being and of animals (the etheric, astral and spirit). Every human being has an individuality, a spirit, but this is not so for the individual animal. Animals have a group soul and a group spirit. Human beings have, in the course of evolution, developed a consciousness of individual self. The fact that marriage between close relatives — as in ancient times — is no longer appropriate is connected with this. However, since individual animals lack this, inbreeding (also called pure breeding) is the best way of attaining a healthy herd.

The breeding of a farm's herd of cows should be pursued exclusively with animals native to the farm. If the cattle herd is raised in this way, it will in the course of time — about thirty years — have homogeneous blood. This is the prerequisite for the group soul of the animals to become fully active within the herd. A farmer, knowing these things, can influence the group soul. Then breeding in the true sense is attained. Nowadays most breeding is crossbreeding, usually by artificial insemination. This method considers only the physical aspect and ignores the more subtle aspects, thus preventing the group soul from working. These animals can only rely on the powers of physical heredity, which are declining today. This decline can be countered by regenerative impulses from the cosmos working through the group soul. True breeding comes about through a connection between the farmer and the group soul. When the animals are kept well — particularly through feeding and pure breeding with a bull from the farm herd — the group soul can again become fully active. Then quite naturally the right herd will develop for the farm involved. The animals are shaped by the nature of the farm, and grow tamer.

The cattle will increase the utilizing of their feed to a degree almost unknown today. Of course they need to be fed according to their nature. Through the four stomachs the cow is able to feed from land that cannot be used for crops. When a herd is consistently fed in this way, this power of digestion increases from generation to generation. In the summer, cattle must be fed on grass and foliage from hedges and woods. In winter they should have hay and straw, and foliage hay from trees and hedges (preferably cut around midsummer's day) as well as beets and fodder kale as moist feed. With this simple and frugal diet, which is in keeping with their nature, cows look noticeably well. Their hide becomes shiny, their eyes grow livelier, more awake and brighter. Their hoofs barely need cutting. Disease and disposition to illness are rare. Horn deformation, a result of inappropriate feeding, disappears. In the course of a few generations the horns grow again in a half-moon form; their substance changes, becoming finer, smoother and shimmering.

The horns enhance the internal play of vital forces in

the cow, contributing to the process of milk development, and are an important factor in improving milk quality. This has been shown by the crystallization method.[9] Milk from dehorned cows is of inferior quality.

Comparing the images on pages 40–41, the first two show healthy conditions as with normal cows. The second two pictures are from dehorned cows on the same farm. The inorganic structures clearly show how the life forces are impaired through dehorning.

The highest quality of milk is achieved with an indigenous herd fed as described above, the feed being exclusively from the farm. Initially after transition to this system, cows will give considerably less milk, because their digestive system and ability to utilize this feed is not sufficiently developed nowadays. These abilities have to be redeveloped by good feeding and breeding. It takes several generations for the first stomach to grow larger and the intestine to grow longer. It is only possible to strengthen the powers of digestion by good feeding and pure breeding. It requires courage and perseverance at first. Milk produced in this manner is very wholesome, and slowly the yield increases, which can be seen as gratitude for the good care: a reciprocal giving and taking that benefits both people and animals.

The rising age of these cows is most impressive. For example, on my farm the average number of calves each cow had was eighteen, and I've never had a single case of infertility.

But of greatest importance is the dung which these animals produce. Used well, it gradually brings fertility, even high fertility, to the most barren soils. Experience shows that a herd kept in this way is the source of health and fertility for the whole farm, including other farm animals who also benefit from the dung. Horses stay healthy eating the grass that grows on cowpats, which the cows avoid. Pigs will eat fresh cow manure, if given, and chickens also eat the dung, to the benefit of their health.

It is painful to see that today almost all cows are fed with 'scientifically' calculated rations, including a large portion of concentrated feed even in summer. In winter they get silage and large amounts of concentrated feed. This has dire consequences for the cow because her digestive system is not suited to such a diet — especially not the rumen (the cow's first stomach), where the feed undergoes a lactic acid fermentation. Then the contents of the rumen are chewed over again, up to 72 times, before moving on through the other stomachs to the intestine. The food substances generally remain for about 36 hours within the animal's organism, being enriched with vitality and fertility. If fermentation in the first stomach is disturbed due to inappropriate feed (grain silage, maize silage), organic poisons are created, particularly hydrogen sulphide, which is highly toxic to the nervous system.

The digestive system of the cow tries to eliminate these poisons quickly, which is possible in three ways. She ruminates less and then gets diarrhea. This shortens the duration in the intestines to about 18 hours. Under these conditions the intestinal content cannot be as enriched with vitality and fertility. Due to the production of hydrogen sulphide, the excretions have an abominable stench. This dung, generally brought out on the fields as liquid manure, immediately kills all life in the soil — it is even worse than artificial fertilizer. Due to this loss of life, the soil no longer has the ability to destroy pathogenic organisms, which now have access through the plants to animals and human beings, thus causing illness. The ability of the soil to retain water also decreases, and pollutants can enter the groundwater unchanged.

The second path of excretion is the milk. Suddenly the cow gives more milk. But this milk cannot really turn properly sour. Instead it turns bitter and spoils after a few days. The proteins have changed and hence are not so healthy

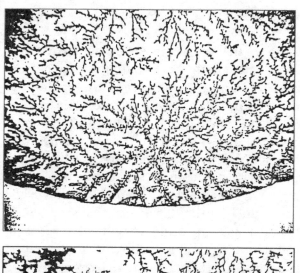

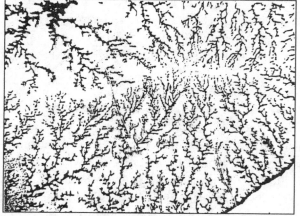

Crystallized milk sample of a cow with horns (enlarged 40 times). These pictures show the periphery which reflects the conditions of the life forces within the nervous-sensory system. Fine, dense, plant-like structures are predominant in both pictures, indicating intense vitality. The harmonious forms indicate a good relationship between the metabolic and the nervous-sensory systems.

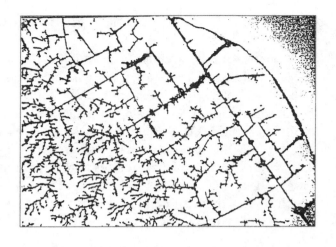

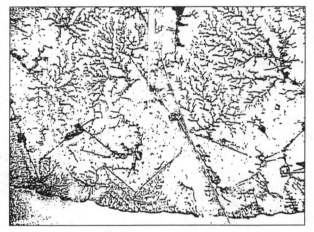

Crystallized milk sample of a dehorned cow (enlarged 40 times). The crystal structures in the periphery are less strongly developed. Rectangular and linear structures dominate. Part of the edge of the upper picture is severed by a long, straight crystal needle. This indicates that a region of the animal's body is not fully integrated. Only towards the centre do plantlike crystal forms appear, expressing health. In the lower picture there is a long, straight needle with rectangular, lifeless crystals near the edge. These crystals indicate metabolic waste products and waste deposits. Linear structures can be seen as an indication of sclerotic processes. Both symptoms appear only in connection with degenerative processes. In human blood crystallization tests, linear crystal needles separating an area from the totality of the picture indicate a disposition for degenerative or even malignant processes.

for humans. One of the consequences of this inferior milk is the enormous increase in allergies, which can generally be traced back to an allergy against milk protein.

The third path of excretion is the development of cysts in the ovaries. This leads to infertility and the cow has to be sold for slaughtering. In Germany the average life span of a dairy cow is two calves. The trend is in the direction of even fewer calves. All these cows are in a latent state of ill health, which makes them susceptible to many diseases that are then treated with chemical remedies, affecting the animals even more.

These experiences lead to the following conclusion. If a change is to be brought about in agriculture as a whole, then it is necessary to start with the cow because she produces the dung necessary for the fertility of the soil, which is a prerequisite for all life.

Unfortunately this will be a slow process since cows no longer exist as pure local breeds. Today in Europe almost all cows come from America or from cross-breeds with such cows. As most of the old, local breeds no longer exist, it will be a long, hard task to get agriculture on the right path again. But we have no other choice.

Raising calves

It is very important that when a calf is born, its first impression is of a human being, of the person who will care for the calf in future. This conditions the calf towards people, and ensures tameness and friendliness. If it is not conditioned in this way, it can fall back into an irreversible wild state.

The first three days are decisive for the animal's health. After being rubbed dry with straw, the calf has to suckle the colostrum milk from its mother's udder within two hours after birth for three days. The colostrum provides antibodies which the calf lacks at birth. After three days the calf is taken from its mother and fed by its caretaker from a bucket with the milk of its own mother and at udder temperature. This is important, for the mother's milk changes in composition daily, being exactly suited to the age of the calf. The separation of the calf from its mother and the feeding by hand develop a bond between the calf and human beings. During the following three months the calf drinks six to eight litres of milk daily in this manner. After this time, the milk is gradually reduced until it gets only one litre daily in the sixth month. A tablespoon of soaked and boiled linseed should be added to the milk, and as soon as possible minced raw carrots should be fed additionally. This is to enhance an early development of the sensory and nervous system. For the reduced milk, the calves can drink water freely from the watering trough.

After the second week, the calf should start eating hay in order to develop its first stomach as early as possible. The hay has to be the best quality. It has to be cut when ripe and processed in the best manner (more about this in 'Meadows and Pastures'). The stall of the calf should be strewn with rye straw, which has by far the highest silica content and hence a good influence on the calf.

The first six months of a calf's life are decisive for its health and longevity. This way of raising calves requires a lot of work and is expensive. But it pays off since calves raised like this are much healthier, have greater vitality and live longer.

As soon as possible in spring the calves are put on a fresh pasture that has been mown in the previous year or been grazed by other animals, for instance, sheep or geese. This protects the calves from parasites.

In summer the calves are fed with grass and in winter with good hay, some rough ground oats, carrots, and then with as much oat straw as they like. Hay from hedge leaves can also be offered. After this first year of intensive tending and careful feeding, in the second year they get used to roughage.

The animals eat grass and leaves from the hedges. When summer pastures are finished, they get only a little hay, some beetroot, but mainly as much straw as they want as well as leaf hay.

When the animals are about two and a quarter years old, they are mated — by no means sooner, for only at this age are they fully mature. Thus they have their first calf when they are about three years old.

These are general guidelines concerning the healthy raising of calves. The main point is the personal connection to the animals, which should be embedded in a loving atmosphere. Later on the animals will thank us in the form of longevity, great fertility and wholesome milk.

Keeping horses

Horses have become less important today, being kept mainly for recreation and sport. Their keeping has become ill balanced and dropped out of the farm organism. In consequence they suffer from parasites. This condition is treated chemically, which in turn leads to a permanent weakening of the horses.

Keeping horses in a healthy way is only possible, if all animals on the farm are in the right proportions and when horses and cows alternate grazing on the pastures.

Formerly, horses were put on a pasture which cows had grazed on before. The horses would eat the grass growing on the cowpats (which the cows will not touch) and thus stayed healthy without the need for chemicals. Only in this way do they attain their full working capacity. All animals in a farm organism depend upon healthy and well-fed cows for their own health and in order to create their own distinct body protein. The healthy protein produced by cows serves as model for other animals. Even people are affected, as already mentioned (we now see increasing allergies to dairy products). The importance of keeping horses on a farm can be seen in connection with their manure, which is of special value for certain plants and should be mixed with the other manure in a certain proportion. The importance of horse manure is in its development of warmth and aroma, which quickly attract earthworms.

Generally speaking, a horse needs one and a half times the feed of a cow. This was one of the main reasons for abandoning the horse. However, it is overlooked that due to their manure, crops gain in quality and quantity to such an extent that horses literally produce their own feed.

For this reason, I had an agreement with a horse farm in that we sent them our two-year-old cattle to graze on their pastures in summer, and in turn got horses from them to graze on our pastures. The cattle gained surprisingly in weight. The pastures improved and the health of the horses was obvious.

It is especially important on heavy, clayish soils to keep horses, to supply the soil with warmth. In such conditions keeping horses should be seriously considered. However, hiring additional labourers to tend the horses can be economically impossible. The great amount of work required from five am to nine pm is unimaginable in our present-day society.

Pigs and their relationship to other animals

Pigs are — as mentioned earlier — waste utilizers. They see to it that all waste on the farm is consumed. This prevents rotting, which produces hydrogen sulphide. You could say that pigs are responsible for the hygiene of the farm.

The amount of waste accumulating on the farm determines the number of pigs. Pig manure is cold and can only be applied in small amounts to heavy soils. Sandy soil, however, can use larger amounts of pig manure to cool it. The aroma of pig manure attracts earthworms, encouraging them to settle in the compost.

Before being slaughtered, pigs have to reach a certain maturity, weighing 150–200 kg (300–450 lb). They achieve this weight by slow growth. The bacon is then of great value for it has certain healing powers. Formerly it was used as a base for ointments for both humans and animals. This healing capacity of bacon is due to cosmic forces which accumulate in the body during slow, natural growth without the use of accelerating feed.

Of course, when pigs are taken out of the natural farm organism and are mass produced, it is not surprising that the original healing power changes into the opposite.

Like cows, pigs should be of a local breed. Ideally they should be raised outside on a pasture with some trees, a 'wood pasture,' with pig huts where they have their piglets. Later the piglets are also fed waste from milk processing, grains and potatoes in an earthenware trough. When they are larger, they should be fed with as many stinging nettles as possible. When they weigh between 30 and 40 kg (65–90 lb), they will be brought into a pigsty, well strewn with rye straw, to be fattened.

The purpose of fattening is to achieve the best ripening process for good bacon and meat. Large amounts of protein as feed do not achieve this because the animals grow too fast, producing unripe proteins (similar to the milk of inappropriately fed cows). These proteins are unhealthy for humans, and physicians have for years been discouraging the consumption of pork, citing sutoxins.

The best feed for fattening is a varied feed from the kitchen and garden, from milk processing, waste from processing grain and, if available, as many nettles as possible. Each pig should also receive cow dung with some grass sod or compost every day. A pig fattened in this way will be of the highest quality and its bacon will have healing qualities.

The importance of sheep

Sheep also form an important part of a farm organism. The atmosphere in a sheep barn is particularly peaceful. Through their wool sheep provide us with warm clothing, blankets, hangings and carpets.

Sheep are ruminants, so their dung is of high value, but it must only be used in homeopathic quantities. An excess leads to soil degeneration. The iron compounds in the humus are dissolved and settle into the subsoil as iron, preventing an exchange between the surface and the depths. This is especially critical to water regulation, causing the soil to dry out. Sheep eat all shrubs and trees right down to the ground, ruining whole landscapes. There are many historical examples showing this — from the Lüneburg Heath and the Balkans to the Scottish Highlands and New Zealand.

Formerly, in Central Europe every farm had a few sheep. In spring a shepherd took out the herd to tend the fields. In fall they were herded with the help of dogs over fields with young grain crops sowed in late summer, to stimulate their growth. Their droppings are important for the soil, and this form of grazing strengthens the grains which develop additional stalks. Sheep also compact the soil with their hooves, impregnating it with certain hormones. Sheep dogs keep the herd slowly moving, otherwise the crops would be grazed right down to the root, hence the saying, 'the sheep has a golden hoof but a poisonous mouth.' For this reason they cannot be herded over the same area for a second time within a year. Keeping sheep properly is a true art. It is possible to considerably enhance the fertility of fields, and even of a whole landscape, or to ruin it all by keeping sheep in a one-sided way.

Heavy soils, being generally cold, can support and benefit from a greater density of sheep than normal. Generally ten sheep on 50 hectares (120 acres) are adequate, but on heavy and cold soils the number of sheep can be doubled or even tripled.

In winter sheep are kept in the barn and fed with hay and some oats. After the last cold spell in spring the sheep are shorn.

In spring, depending on the breed, the ewes have one or two lambs. The lambs stay with their mothers, feed on their milk and soon start eating grass. So raising lambs is no problem.

Goats, the gourmets

Goats, also ruminants, are gourmets. They are hard to keep, because they will very quickly nibble away at any shrubs, flowers, buds and the vegetable patch, causing much damage.

If goats are kept, they should live in the cow barn, which is traditionally believed to benefit the health of the cows. Often goats are kept because their milk is needed for therapeutic reasons. The kids are raised as easily as lambs. Several kinds of valuable and healthy cheeses can be made from goats' milk. However, keeping goats is similar to holding sheep. Kept in small numbers, they enhance soil fertility, in greater numbers they are destructive. Goats were originally mountain animals. They were domesticated by breeding, and it became possible to also raise them on flat land.

Chickens, domesticated birds

Chickens represent the bird world in the farm organism and used to live in a loft, directly above the cows. Chickens and cows formed a kind of symbiosis. In spring the chickens hatched there, having been incubated by the brooding hen for three weeks. The warmth of the cows rose up, and the floor was always warm for the chickens. The breath of the cows, drifting upwards, enveloped them. During the day the chickens jumped down, feeding between the cows. They also ate some fresh cow dung, which kept them healthy, enhancing development of proteins in the eggs. Today, generally, this is no longer possible. In summer chickens ran about in the orchard and around the farmyard, clearing the area of vermin and waste, in a kind of competition with the pigs. They were fed with waste grain (small or broken grains) from the cleaning machine. Old tradition advises to keep only two chickens per hectare (one per acre) of farmed land. Here again the droppings play an important role (phosphorus effect) and should not exceed a homeopathic dose. It is possible to make liquid manure with chicken droppings, which is very helpful for broccoli, cauliflower and flowers in general.

It is good to keep some guinea fowl along with the chickens. Their shrill cries chase the rats away. Unfortunately guinea fowls' nests are hard to find; their eggs are a delicacy. They look after their offspring themselves without assistance.

Geese, the guardians

Geese once saved the Capitol in Rome with their screeching. They behave like this on the farm, and every stranger will be greeted with loud screeches so no one goes by undetected. They are an alarm bell.

Geese are also part of the bird world of a farm organism. They live in symbiosis with sheep, who will even eat grass growing on goose manure, which no other animal will touch.

Generally two geese and one gander are kept for the long term. In early spring they start laying eggs (up to sixteen eggs per goose). If their eggs are removed, they will lay many more (up to sixty per goose, depending on the breed). The eggs have to be hatched in an incubator.

When the goslings hatch, they are fed with barley bran and stinging nettles, primarily with chopped nettle leaves.

Most breeds of geese today do not have a dependable brooding instinct. To strengthen this instinct, stinging nettles can be put under their nest. Good geese can grow very old, so it is better not to slaughter them but to continue to use them for breeding.

As soon as the goslings develop their feathers, they can be put out to pasture together with the sheep. There the geese kill off parasites and vermin. This helps sheep, who are especially susceptible to parasites. It is important to see that geese always have enough young clover grass (mixture of grass and clover) and enough water — at best a pond. In fall, oat or barley bran can be fed additionally before the fully-grown geese are slaughtered. Goose down and small feathers can be used for pillows, quilts and bedding. In fall, when geese change their feathers, it is possible to carefully pluck the down from the stock of breeding geese before they fall out — as it was done in olden times.

Sheep and geese should have several pastures — four if possible — to change pastures frequently. In the following year calves and cows can use these pastures, so that sheep and geese benefit from the cow dung by annual rotation.

Ducks

Ducks should also be present in a sound farm organism. Their primary task is to keep waters, pastures and meadows free from parasites, which they do very effectively. They need a fairly large pond. For raising them, it is best to have a drake and two to three ducks. After the ducklings have hatched in spring, the mother ducks wander off immediately with their offspring to the pond, streams and ditches where they find plenty of food. Later, when the young ones have grown a bit, they swarm out onto the pastures freeing them from pests. Ducks should not receive much additional feed, otherwise they will not be keen enough to devour pests. They should only be fed some bran in fall in order to gain sufficient weight. The superfluous ducks can then be slaughtered and sold, keeping only the ducks used for breeding. In the evening, both ducks and geese have to be locked in a hut by the pond or in a stable well supplied with straw, so that foxes have no chance of a tasty meal during the night.

Cats

The farm must have cats. They catch mice and the bigger ones also catch rats. You just have to ensure that not too many kittens are born, because if there are too many they will go for birds. Experience has shown that it is good to offer cats a small dish of fresh milk in the morning and evening to enhance their mouse hunting.

The farm dog

The farm must have a dog, which needs to be familiar with the entire farmyard and the surrounding territory. He has to report every irregularity in the area immediately. The people caring for him should treat him with special affection so that he develops protective instincts for the whole farm. The dog is a kind of extended sense organ of the people living on the farm. He has to be selected with special care, ideally born on the farm.

Wildlife

The world of insects

The world of insects calls for special attention. The most important insects are bees and ants — especially the red ant. These creatures are very sensitive to radiation of all kinds, and are more and more endangered in our modern world. There are large forest areas now where no anthill can be found. Ants — as well as bees — produce formic acid, which they distribute across the landscape. Bees leave a trace of their poison, formic acid, when flying to flowers to collect honey. This strengthens the life forces of the plants, raising their yields considerably.

Ants do the same with their formic acid on the soil, in hedges and forests, where they collect pests and raise aphids on trees. The aphids live in symbiosis with the trees. Bees also gather the liquid excreted by aphids, which they transform into forest honey. This complex interaction permeates the whole of nature, right into the atmosphere, with formic acid.

Formic acid is the bearer of life forces for the whole of nature. Everything — even the human being — has to be permeated with formic acid, although humans can synthesize it from oxalic acid in their food. Hence it is vital that we create the right living conditions for these insects. The landscape has to be permeated with flowering hedges and with woods, and appropriate places for beehives have to be found. Like ants, bees also orientate themselves by terrestrial radiation.

Ants generally come by themselves, and settle over nodes of underground watercourses. However, since the whole atmosphere today is permeated by electromagnetic radiation, they will often not manage to settle without assistance; they become irritated. We have to consciously help them. Planting as many hazel bushes in hedges and woods as possible helps to neutralize ambient radiation. The whole landscape should be sprayed regularly, at least once a year (better twice) with horn silica and horn manure biodynamic preparations (see p. 75). This neutralizes the effect of electro-smog more thoroughly than hazel bushes, and ants will settle again. Only then will good humus be formed and fertility greatly increased. Due to bees, epiphyllic yeast develops on the grass. When cows graze, this yeast enters their first stomach, creating optimal fermentation. In this way all cultivated areas receive the right dung, which boosts the fertility of the soil. The activity of the insect world can be seen as one of the foundations of a living agricultural organism.

Butterflies also play an important role in the relationship between nature and cosmos. So we have to ensure appropriate conditions for them. Stinging nettles provide the appropriate habitat for a variety of butterflies.

The importance of birds

The conditions for a diverse and plentiful bird population are created by planting an extensive network of hedges and laying out ponds. The more diverse the shrubs and trees in a hedge, the greater the diversity of birds — for every species has its own connection to a particular kind of tree or shrub. It is important to have many thorny bushes in hedges, like hawthorn, blackthorn and wild roses, because they provide birds with a protective habitat from cats and birds of prey. It is also good to put a variety of nesting boxes on trees and buildings.

As well as being great devourers of pests, birds gather the astrality (soul forces) collected by trees and shrubs from the air, and spread this to the neighbouring crops; every plant needs to be enveloped in astrality. In the Agricultural Course, Rudolf Steiner indicated that astral forces carry nitrogen. Birds accomplish this task by flying over fields, as well as with their singing, which can rise to a true concert. These concerts begin with the germination of plants in spring and foster the growth of all vegetation; they cease around midsummer when the ripening process begins. Recent research has shown that plants react strongly to music: classical music promotes their growth, while rock music inhibits it.[10] Something of this interaction is echoed in the opening words of St John's Gospel: in the beginning there was the Word (sound, tone). The patterns produced by Chladni plates demonstrate the structuring powers of sound and different musical notes.

In summary, we can say that birds are necessary for the healthy growth of plants, as the alert observer will readily notice. Scientific research concerning this matter already exists.[11] Here too, as with insects, the whole area needs to be treated with horn manure and horn silica preparations. The horn silica preparation in particular has an effect high up in the atmosphere, influencing the environment of birds positively.

Bacteria, earthworms and moles

We have looked at the animal world living on the surface of the earth, with its complex interaction. But there is also intense life within the soil which, in symbiosis with fungi, algae and others, serves the development of healthy and fertile humus. Humus is the bearer of life forces, which in turn are the prerequisite for healthy plant life. Healthy soil life has to be fed and nurtured. Stubble on fields, extensive root systems and organic fertilizer — such as well-manured compost which soil incorporates immediately, turning it into humus — serve this purpose. Soil life is also stimulated and promoted by horn manure spraying, the spraying of valerian, stinging nettle tea, or an extract from rotten stinging nettles or other herbs.

In one gram of well-tended soil there are:

2.5 billion bacteria

700 billion actinomycetes

400 billion fungi

50 billion algae

30 billion protozoa

These numbers give an impression of the complex structure of healthy humus soil.

There are also a large number of worms in soil. Earthworms — there are many different kinds — are truly miraculous in their efficiency. Every day they eat a quantity of earth, with its organic waste, equal to their own body weight. This matter is transformed into humus by digestion and deposited on the surface of the earth as worm casts. These casts have repeatedly been analysed for their nutrient factors and, in relation to the surrounding soil, have been found to contain:

5 times more nitrate

7 times more phosphorus

11 times more potassium

2.5 times more magnesium
2 times more calcium

This wonderful ability of earthworms contributes greatly to the fertility of the soil. About two hundred earthworms per square metre are desirable, or seventy to eighty worms for every ten paces of ploughed furrow.

Their activity in the ground, burrowing many tunnels to a depth of one to two metres, creates optimal loosening and airing of the soil. This prevents putrefaction, and after a heavy rainfall water easily soaks away to join the groundwater, thus preventing erosion.

The worm is also an untiring assistant in composting. The common redworm, Eisenia foetida, comes in such large numbers and with such an appetite that is often unnecessary to turn the compost pile.

Another relatively large subterranean creature is the mole. Often regarded as a pest because of molehills, close observation, however, shows that wherever moles have burrowed, growth improves. They loosen and aerate the ground and molehills become enriched with etheric life forces. After levelling molehills and spreading their soil, plants (grasses) grow better and additional fertilizing is often unnecessary. As well as horizontal tunnels, moles dig sloping tunnels which help drain and regulate the ground. So the mole is also important for the healthy development of soil.

Such soil is full of life and is of great hygienic importance within the totality of nature. It prevents germs and harmful development from the start and is a source of health for all of nature.

Fields and pastures

Meadows and pastures, the bearers of fertility

An old saying says the pasture is the mother of the arable field. This can be understood in the sense that through tending pastures in the right way, using all the animals of a farm, an excess of etheric life forces is created. This nourishes other parts, especially the field crops. Natural meadows and pastures are most valuable, for over the course of decades or even centuries, the right relationship between grasses, herbs and clover has come about, together with good, appropriate soil life. The quality of plants given by nature influences the animals — particularly the cows — eating them. The dung then has a special quality which is very good for pastures as well as fields.

It is possible to increase the quality of pastures, which affects the animals feeding on them. Their dung then becomes better, creating an ascending spiral of fertility that brings about a slow but continual increase in quality over decades. This slowly increasing quality benefits the whole farm organism, especially arable soil.

Because it takes eighteen years for the original pasture quality to be attained again after the land has been tilled, it was formerly regarded as sinful folly to plough up old pastures for temporary tillage.

Pastures have a self-induced fertility due to the diverse kinds of clover, and the way the pastures are used also influences the development of the plant community.

The first measure to be taken in early spring, before grass starts growing, is to spray pastures with horn manure preparation. Then go over it with a harrow. These activities should be done on leaf days and when the moon is descending (see The Biodynamic Sowing and Planting Calendar by Maria Thun). The second measure to be taken is the application of horn silica — also on leaf days, when the grass has developed sufficiently. This should be done in the very early morning hours.

Very young, 'unripe' grass in early spring is not good for cattle. A large part of pastures can be used for making hay. For good quality hay, protein and silica in the grass have to be fully developed, which is generally the case around midsummer's day in June. By that time, diverse herbs have grown, which would be lacking in hay if the meadows were mown earlier. If mown too soon, proteins are still developing, and raw proteins (nitrogen bearing substances) create disorder in the cow's first stomach. By around midsummer's day, the raw fibre in grass blades has developed sufficient silica. Silica, as the bearer of the forces of light, is needed for digestion in the cow's first stomach.

Mowing around midsummer's day is in harmony with nature: fawns are able to run away when the mowing machine approaches, and young birds can fly away from their ground nests. It is good to mow when the moon is in a light or warmth constellation, like Gemini, Libra and Aquarius, or Aries, Leo and Sagittarius. Mowing itself should be done early (four to six am) because grass is still damp with dew at that time and can be mown more easily, and strong forces of the sun from the constellation Pisces flow to earth and influence the grass. Grass should be cut cleanly, close to the ground. This will make it grow thicker and the sod will become firmer. A clean cut stimulates growth, giving a greater yield at the second cutting.

A mower with a double knife system is best, and it cuts a wide swathe. Take care to adjust the cutter well and ensure the blade is sharp. While they need more maintenance, these cutters save on fuel. Circular cutters do not cut the grass but break it off the stalk, causing a slower regrowth. They also need more fuel. Bees and other insects on the grass will be killed by circular mowers but not by the other types.

After mowing, spread the grass immediately and turn it over once during the day. In the evening, rake it into rows. At night a slight fermentation takes place. In this way hay is protected against the dew falling at night. In the morning, after the dew has evaporated, the hay, now half dry, should be spread out again. Then turn it over twice that day, but more slowly and carefully than before. In warm, sunny weather it can be dry by the late afternoon. It can then be gathered into long rows, from which it can be loaded onto a cart or baled. Letting the hay dry on the ground has the advantage that it is exposed to a lot of light. The disadvantage is the risk of inclement weather. However, if it is raining, hay can be put on drying racks. Leave it on the racks until totally dry. While this method avoids risk, it is a lot of work. Another disadvantage is that hay on racks cannot assimilate light forces as well as when it dries on the ground.

Use and care of pastures

Putting cows to pasture in the spring takes careful preparation, observing the animals closely. In early spring, when grass is sprouting, it is possible to herd cows quickly over grasslands, letting them eat the tops of the grass. It is as though this first grass has a medicinal effect on the cows. Their hide grows more beautiful and their eyes shine. In turn, through this first grazing, the grass will be stimulated to thicker growth. After this very first grazing the regular pasturing of the cattle should not start too soon, as the vigorous grass initially has too much nitrogen. It is better to keep feeding them hay and straw during this time, only letting them onto pasture later. By doing so, this critical transition is made easier. Later the cattle will be out day and night, only being brought in for milking in the morning and evening.

Cows much prefer to graze in the half shade, avoiding glaring sun. Surrounding the pastures with hedges, and leaving large, solitary trees, gives shade as well as foliage for feeding. Ideally there should be several pastures. When one is grazed down, the cattle are put onto the next pasture to allow the first one to regrow. Treat grazed pasture with horn manure preparation (see p. 75, 77) and then harrow it to level out any molehills, to spread the cattle dung and to air the sod. This procedure should be carried out on every grazed pasture.

When the grass has regrown, treat it in the early morning hours of leaf days with horn silica. After the end of the hay season, grass continues its growth. The different kinds of clover in particular grow better now because they need more light for their growth. Generally a second cut of hay is possible in early fall. If not, the mown grassland can be used as pasture. Make sure that pastures are not grazed too short as winter approaches. In late fall, treat pastures again with horn manure. Despite about a third of the ground being covered by cow dung — and one could believe this to be adequate — additional compost should be spread to stimulate humus and soil development. It is necessary to prepare this soil compost with special care.

Meadows

Places where the water table is higher and which tend to be damp are suitable for making meadows. Isolated trees and shrubs of different kinds can grow here. Mushrooms will readily appear, and this helps to regulate parasites. Mushrooms have a particular relationship to this class of creature, preventing their harm from reaching other forms of life. Mushrooms perform a similar task with parasites as coniferous trees do with birds, preventing them from harming the farm organism.

Tilling the fields: the great art of farming

The aim of farming is to increase the fertility of the soil. So each farm is an individual entity, depending on the type of soil, the climate and the farmer's powers of observation. None the less, there are general, basic principles in farming which begin with the question, Should I plough or should I work without ploughing?

The concept of farming without ploughing is based on the idea that the natural structure of the soil should not be disturbed and destroyed by the rough interference of the plough. Other tools have been developed which enable ploughless farming. However, experience has shown that ploughing gives better yields. This is due to plant growth triggering the disintegration of humus, which does not occur in unploughed soil.

In all this we must bear in mind that the aim of farming is to increase the fertility of the land, to infuse it with more life. The ancient Persians knew this when they introduced ploughing to cultivate plants, which the earth would not bring forth by itself. They opened the soil using a hook, enabling cosmic forces to permeate the soil. Through these cosmic forces our present-day cultivated plants, our food, was developed. However, this ancient ploughing did not turn the soil, but simply tore it open, mixing the uppermost parts. Gradually this form of tillage was developed and refined, but it never consisted of turning the soil, only mixing it. This work was always done in harmony with the sun, moon and planets to ensure the cosmic forces would be absorbed by the earth. This needs to be the basis of our agriculture today, whose sole aim should be to enliven the soil.

The following can help bring this about. After the grain harvest, fields should be treated with horn manure before doing anything else with them. Then a thin layer of well-prepared manure compost should be spread on them. Then the uppermost soil (2 cm, 1 in) should be turned with a light plough, just covering the manure with soil. Water rises from below by capillary action when the sun warms this soil covering. Underneath the covering it becomes warm and damp, encouraging rapid bacteria growth and activity, which is controlled by the layer of manure. Nine days later, these fields need to be ploughed again, but now with a different plough, approximately 10 cm (4 in) deep. The plough should not turn the soil, but only mix it well. To do this, the ploughshares need to be almost vertical. After the soil life has reached deeper strata, nine days after the second ploughing, the fields can be ploughed again with the same plough, now to the full depth of the humus layer, but no deeper. If even a little subsoil is mixed in, the positive soil condition attained collapses and all the work would be in vain.

This rhythmic working of the soil over periods of nine days together with the manure causes fermentation within the humus stratum. This fermentation process makes the soil especially receptive to cosmic forces. For this reason this work should be done when the sun is in the constellation of Virgo, while the moon is descending and ideally in the constellation relating to the new crop. All work must be done slowly, for working the soil quickly disturbs humus and its conditioning. Ploughing should not be faster than a horse walks, about 5–6 kph (3–4 mph).

After this treatment, and after the soil has been allowed to settle for a few weeks, the soil should be harrowed and treated with horn manure preparation. Then the seeds germinating in winter can be sown.

For spring sowing, soil can remain in rough furrows. It will collect cosmic forces which radiate down upon the earth during the whole winter. Generally the more the soil is moved by hoeing, harrowing, furrowing and ridging, the more cosmic forces it gathers.

The aim must always be to take measures that will bring more life to the soil. This means that as many plants as possible — potatoes, beets, cabbages and all sorts of vegetables — should be grown on raised beds. It is especially effective if, after the soil is treated with horn manure, it is ploughed, harrowed, spread with well-rotted dung, and then ridges are made with the ridging plough, banking the earth to the right and left.

In olden times, all fields and even pastures were ploughed to form ridges (ridge and furrow), for people had an instinctive knowledge that soil raised above general ground level would be filled with vitality. This can even be observed with molehills whose soil shows greater fertility. In my experience, healthy plants will grow in soil that has been enlivened in this manner, and will contain all the necessary minerals, vitamins and trace elements. Some minerals and trace elements even appear in the soil benefitting the crop following. In principle, it can be said that plants measurably enrich the soil's components. This happens in the following manner.

Dung increases the vitality of soil so that plants are supported in their growth by the soil's life forces. Soils lacking in life forces, having been treated only with mineral fertilizers, bring forth plants that contain few essential vitamins and trace minerals. Hence they are not sufficiently nourishing for either humans or animals. There is a continual exchange between soil and plants, a continual giving and taking with transmutation of substances. Rudolf Steiner mentioned this in the Agricultural Course; this was first suggested by Baron von Herzeele, and then verified by Rudolf Hauschka and C. Louis Kervran. We have to learn to look at the earth not in isolation but together with the whole universe — for our world originated from the universe and, together with the planetary system and the zodiac, forms one vast organism showing interdependency right into the smallest parts.

Vegetables grown only with mineral fertilizers show a decrease of vital minerals:[12]

String beans: 100% loss of sodium, which is important for the nervous system and building up muscles.

Broccoli: 75% loss of calcium, which is important for teeth and bones.

Carrots: 75% loss of magnesium, which protects against asthma, kidney stones (renal calculi) and heart conditions.

Spinach: 60% loss of iron.

The right time for harvesting grain

During the 1950s, grain harvesting changed rapidly with the introduction of the combine harvester. Combines made life much easier for the farmer, but grain quality and soil structure suffered as a consequence, for grain had to be harvested when it was completely ripe or 'dead ripe.' This meant that the process of ripening was now completed on the field, and the further stages that had occurred in traditional harvesting methods no longer took place. The term 'dead ripe' points quite rightly to a rupture of life forces.

Traditionally, grain was harvested about fourteen days earlier, at the stage of 'dough ripeness.' This was better for both grain and soil. Legumes were sown in the harvested fields and their growth during these extra two weeks of summer had a beneficial effect on the soil and humus, which carried on through the rest of summer and autumn. Nitrogen, calcium, phosphorus and potassium increased, as well as there being an abundance of earthworms. Later harvests made it too late in the season to sow legumes, preventing the amelioration of the soil. The soil has slowly deteriorated more and more. This was not taken seriously at first, as decreasing yields could be boosted by the application of chemical fertilizers.

Following this change in harvesting practices, it was also found that the germination capacity, vitamin and trace mineral contents of grain had declined steadily. Nowadays, grain has a germination capacity of only two years. In the second year the power of growth is distinctly reduced. In the eighteenth century, rye was known to have a germination capacity of 150 years. This shows the importance of preserving the vitality of grain, and this does not happen when it is harvested at the 'dead ripe' stage with a combine harvester.

To harvest a cereal crop with all the vitality and etheric

forces needed for our health, we have to look towards the older methods of harvesting. For the health of an organism — be it plant, animal or human — the intensity of cell respiration is decisive. Cell respiration depends on the vitality and activity of soil, and is noticeable in the activity of mitochondria. The less activity there is in soil, the fewer the mitochondria. Mitochondria are the 'power plants' of living cells and their enzymes help maintain vitality in grain. They activate a process of cellular respiration and also serve genetic functions. There has been a worldwide reduction of mitochondria in soil, showing that etheric life forces and vitality are diminishing. Dying soils affect the vitality of plants, animals and humans, leading to a steady weakening, which in turn makes organisms more prone to disease.

This again points to the damaging effects of current agricultural practice, and calls for a fundamental change in methods.

Treatment of grain during harvest

Grain should be harvested when 'dough ripe' (about fourteen days before the 'dead ripe' stage) with a reaper-binder. The sheaves are placed in conical stooks on the field and left there for about nine days. The slow drying of the stems allows the grain to absorb vitality from them as they gradually die back.

After nine days, the sheaves are taken into the barn where they are built into stacks. Further ripening takes place in the corn stacks due to a small rise in temperature and the increased activity of mitochondria. This continues well into the winter when grain is threshed and spread out in the loft. About every two weeks it is turned.

This way of treating grain preserves vitality that is then passed on into our baked and cooked food.

Seed raising

In order to stimulate these new plant qualities, it is important that farms produce their own seed, as was customary in former centuries. In future, seed production will be of the utmost importance. It is no longer sufficient to produce seed as has been done up to now. Since the nineteenth century, when science was applied to agriculture, specialized farms and big business have done incalculable damage. In future, healthy seed production will only be possible in a well-functioning farm organism, where the necessary vitality and life forces are present. Only then will seeds have the inner forces and fertility needed for the future. These inner forces are dependent on the harmony of all functions in a farm organism, in the world of insects, birds, animals and plants, as well as on the cosmic forces working through biodynamic plant preparations.

These are the preconditions for producing seeds. Additionally, a study of the forces coming from the cosmos at various seasons is necessary. Those of spring are very different from those of the fall or winter. These forces have an effect on seeds. For instance, when sowing in summer, seed can give a higher yield, while if sown later, towards winter, its reproductive power may be enhanced. The greatest effect can be seen when sowing during the thirteen Holy Nights of Christmas. Seed sewn during these days — if possible when the moon is in the appropriate constellation — produces the greatest power of regeneration. The time of day is also important. The strongest growth forces come to earth from Pisces between four and six in the morning, or from Virgo between four and six in the afternoon.

Additional methods of stimulating seed's regenerative power can also be used. For instance, in former times people sang over plants, or made a loam broth and sang while stirring continually; the broth was then sprinkled over plants. Another possibility is with smells: essential oils (not synthetic oils) of rose, rosemary, lavender, etc. can be used.

With these measures, the quality of seed can be greatly improved. Of course, seeds have to be found that have a sensitivity to cosmic effects. Common, selectively-bred seeds do not have the quality needed for such seed development. Older types of seed can be treated in the above manner.

Woodlands, centre of many landscapes

A woodland belongs to every agricultural organism. On a larger scale, forests regulate the climate, while locally woodlands regulate groundwater. They prevent the lowering of the water table, even though on hot days very large amounts of water evaporate from trees into the atmosphere, thus ensuring sufficient humidity. A healthy woodland is mixed. Ideally, it should consist of trees in three stages — young, middle aged and old — growing together. The edge of a woodland should have sufficient shrubs, such as elderberry, bird cherry, willow, black alder, dogwood, buckthorn, rowan trees, hawthorn, hazel and wild rose. This closes off the woodland, keeping air movement out of the interior of the woods.

As already described, ants as well as other forest animals such as wild boar and roe deer in the right numbers are essential to the development of good woodland humus. In order to minimize forest damage, like the eating of young trees and rooting the ground, too many animals must be avoided. The dung of these animals stimulates optimal development of the forest humus. In past times, even cows were herded through woodlands. On one hand this was good for the cows, as their diet was augmented by the foliage they ate, and on the other hand the cow dung bettered the quality of the humus.

Nowadays, woodlands need to be protected from electromagnetic radiation by planting many hazel trees. Biodynamic spray preparations can also be used. Tracks need to be made at regular intervals to allow both the transport of timber and spraying. Spraying will also enhance the conditions for ants so they can carry out their beneficial activities.

Forests and woodland always contain trees compatible with the soil and the climate. When planting new woodlands there is a certain sequence of trees. On more sandy soils pine grows first, preparing the soil for followers such as oak which has deep roots, or beech where there is loam. Larch, a fast-growing tree, can also follow pine. Ash and maple can join where the soil is better. In damper areas fast-growing poplars will appear, and along ponds and riverbanks many different kinds of willows will take root. Birch trees are at home almost everywhere. Lime (linden) trees need better soil. While the elm has become rare, it appears to be recovering. In wet areas and along riverbanks grey and red alders like to grow. On sandy ground other coniferous trees like firs will grow along with pine trees. The fir must be treated cautiously for it makes the soil acid. If planted as monoculture during a storm, wide areas are often totally blown down.

In the long run there will be problems in planting non-native trees due to environmental conditions becoming increasingly difficult. Weakening life forces allow fungal infections and attacks by pests, which can lead to trees dying. Non-native trees have already been weakened by being removed from their native environment and having to grow in surroundings that are not always compatible.

A healthy forest is always mixed, and all plants and trees support each other. The light shines down to the ground so that other plants such as blueberries, cranberries, raspberries and blackberries can prosper. Smaller trees such as holly, bird cherry and wild apple will also grow there.

Woodland kept in this manner will, within a few years, show a good increase of growth and better health.

Compost and Manure

Manure and its treatment

Most of the manure on a farm is from cattle. This dung should contain as much straw as possible derived from the bedding of the animals. If possible, rye straw should be used for its high silica content. The dung from the other farm animals should be mixed with cattle dung. Cattle dung contains etheric and astral forces, as well as the spiritual forces which radiate from the surroundings into the cow. No other animal has this ability. This is why even today the cow is regarded as sacred in India. The development of soil, plants, animals and human beings depends on this singularly fertile cow dung. So it is important to treat the fresh dung carefully to preserve its special qualities.

Before removing the manure from the barn, stinging nettle preparation should be strewn over it. This preparation starts the right rotting process, while preserving all the forces in the dung. The dung together with straw is then carefully built up in layers. Every layer should receive an ample portion of loam broth. Ground basalt can also be mixed in. While being built up, the whole manure heap is treated with compost preparations in the following manner. Make several holes about 50 cm (20 in) deep with a stick. A pinch (as much as can be held between thumb and two fingers) of one of the preparations is put into each hole, which must then be well closed with manure. Depending on the size of the heap, the distance between the preparations should be 50–100 cm (20–40 in). After the manure heap is finished, cover it with soil and treat it once more with the preparations in the above manner, including the valerian preparation. To do this, a teaspoon of valerian preparation is stirred intensively for 5 to 10 minutes in 10 litres (2½ US gallons) of warmed water. It is then poured over the manure heap using a watering can with a fine nozzle. If the heap was started in April or May, it is then left undisturbed until it is needed in the summer. In June or July take it to the fields and again build it into a heap. Treat it with the preparations again and cover it with straw. After three weeks the pile is full of fungus. Now is the time to spread it on the freshly-mown stubble fields. This method of treating dung is the result of extensive experience and observation. Preparation of the barn manure to the point of its application has to be done with great care. A feeling for timing and method will gradually develop.

Compost preparation

For pastures and meadows, high quality compost has to be made. Any organic material from the farm can be used: fall foliage, bad hay, straw, excavation material from ditches and ponds, dead animals, scraps from butchering, garden and kitchen waste (if not used for feed), as well as waste from crops such as potato leaves after harvest, etc. It would also be good to add some pig and chicken manure. Build these diverse materials and manure in layers with lots of soil between them. Lightly strew some quicklime over each layer. When finished, treat the compost heap with the preparations as described above for manure. Then cover with straw, bad hay or dried leaves. Let it ripen for several months.

Then it has to be turned and rebuilt, again treated with preparations and covered with straw or the like. Generally it takes a year for this compost to mature.

The biodynamic preparations

In the Agricultural Course Rudolf Steiner advocated the use of the following preparations in order to counteract the steadily declining forces of the earth.

1) Horn manure: This preparation should be applied to the soil before every sowing. This preparation brings etheric and astral forces to the soil, enlivening it.

2) Horn silica: This liquid preparation is made of very finely ground quartz. It is sprayed on young plants, promoting their growth primarily by intensifying and harmonizing the effects of light on plants. It also brings about a subtle change in the atmosphere, which attracts insects, butterflies and birds, creating better conditions for them.

3) Yarrow has a particular relationship to sulphur, in that it is able to regulate the sulphur processes of the soil. Yarrow makes it possible, through dung, for soil to become sensitive enough to gather substances from its surroundings — for instance silicic acid. The silicic acid content of plants is rapidly declining (it has halved in the last 25 years). Leaves assimilate silicic acid from the surrounding atmosphere. This is shown by plants growing in swamps where the soil does not contain silicic acid. Yarrow also has the ability to regulate the potassium process, ensuring plants always have enough potassium, creating enough when needed during their growth.

4) Camomile also contains sulphur, but has a different quality from yarrow. Camomile primarily works on calcium.

Dung treated with camomile increases in nitrogen content, and has a stimulating effect on plant growth. Plants become healthier due to this preparation.

5) Stinging nettle preparation is of the utmost importance for agriculture and particularly for dung. It regulates sulphur, potassium and calcium. The nettle also gives out an iron 'radiation' equivalent to that of human blood. When the nettle preparation is added to dung, the dung becomes 'sensible,' in the sense of losing the possibility of developing wrongly. The dung transmits this property to the soil. The soil then adapts better to the plants on it.

The nettle can be used in other ways, for instance, as tea sprayed on plants in a dilution of 1:10. A 24-hour extract can be made. Half a barrel of nettles is filled with water. Leave the nettles in the water for a day. The extract can then be used for plant spraying. Similarly, fermented nettle water can be made by leaving the nettles in the barrel for three weeks to complete the fermentation process. The fermented extract has to be diluted 1:10 and sprayed very finely onto plants. This stimulates plant growth.

6) Oak bark preparation is a preventative of plant diseases. To prevent these, calcium has to be administered to the soil, but this calcium must be from a living process. The bark from living oak trees contains 70–80 % calcium in its ash.

7) Dandelion preparation: Today the earth is unable to supply plants with sufficient silicic acid. As previously mentioned, plants only have half the silicic acid content which they had 25 years ago. Dandelion has the ability to assimilate silicic acid from the surroundings with its leaves. Adding the dandelion preparation to dung gives the soil the ability to collect and assimilate silicic acid.

8) Valerian flower extract is highly diluted (1 teaspoon to 10

litres, 2½ gallons of water) and then sprayed over dung. The valerian preparation has the ability to stimulate phosphorus processes in the dung and later in the soil.

The addition of these plant preparations allows the plants to develop greater sensitivity towards their surroundings, enabling them to utilize the substances and forces of the wider environment.

The production of the preparations

Preparation 500: Horn manure

To make the horn manure preparation, collect cow horns (best from cows having had calves, not from bulls) preferably from the local region. At the end of September, fill these horns with good, fresh cow dung and then bury them about 20–30 cm (8–12 in) deep in the soil. Mark the area with four sticks to help find the horns in spring when as many as are needed are dug out. The manure, now changed, is taken out of the horns and diluted with slightly warmed rain or pond water, about 15 litres (4 gallons) per hornful. This is enough for ¼ hectare, so four horns are needed for 1 hectare (2½ acres). Larger areas need less. The horn manure liquid must be stirred for an hour. Stir intensively in one direction until a deep vortex is created. Then quickly change the direction of stirring, creating chaos, until the next vortex appears. Keep changing direction in this way. This stirring results in a total permeation of the water with the vital forces of the manure. The treatment remains in the realm of life, concentrating life forces. Before using it, strain the liquid through a fine sieve or cloth to filter out solid particles so that the spray nozzle does not clog. Then pour the filtered liquid into a portable back sprayer or — for large areas — into a tractor sprayer. About

60 litres/hectare of horn manure is sprayed before sowing (6½ gallons/acre). Spraying should be done in the late afternoon, preferably during the descending moon.

The horns not dug out in spring can stay in the ground until they are needed.

Preparation 501: Horn silica

The horn silica preparation is termed preparation 501. Ground quartz crystals (quartz, orthoclase or feldspar) are used. It is important that it be very finely ground to the consistency of flour. Generally one hornful is enough for a larger farm (100 hectares, 250 acres) for several sprayings since only a very small amount is needed for each spraying.

Make a thick paste with the quartz powder and fill up a horn. Bury the horn in the humus layer of the soil (20–30 cm, 8–12 in deep) in spring, and leave it in the ground throughout summer. At the end of September, at Michaelmas, take the horn out of the ground. The contents can be stored indefinitely in a glass. Only a very small amount is needed. Take ½ teaspoon to 60 litres (16 US gallons) of water and stir for an hour in the same way as described above for horn manure.

It should be sprayed on young grass, shrubs, fruit trees and flowers in the very early morning hours, on a sunny day. Use a very fine nozzle to create a mist. About 60 litres is sufficient for 1 hectare (6½ US gallons/acre).

Preparation 502: Yarrow

In early summer, pick the little white flowers of yarrow (Achilea millefolium) in the morning. Put them out to wilt, but not to dry totally. When wilted, put them in a dried stag's bladder. (The bladder must be dried so it does not tear.) The bladder is then hung in a sunny place for the whole summer. In fall, around Michaelmas, bury the bladder in good soil at a depth of about 25–30 cm (10–12 in). In spring, around Easter, dig it up, take the contents out of the bladder and store them in an earthenware pot. Keep the pot in a cool cellar in a box surrounded with peat.

Preparation 503: Camomile

Pick camomile flowers (only Camomilla officinalis) preferably in the mornings and dry them. In fall the dried flowers are moistened with camomile tea and then stuffed into bovine intestines. About 30 cm (12 in) length is sufficient for a farm. Tie the ends securely. In late September, at Michaelmas, bury these 'sausages' about 25–30 cm (10–12 in) deep in ground where the snow stays the longest, yet which gets a lot of sun. Then in spring, shortly after Easter, dig it up, taking it out of the intestines, and treat like the yarrow preparation 502.

Preparation 504: Stinging nettle

At the latest point the nettles are still flowering, mow a sufficiently large number of nettles (Urtica dioica) and leave them to wilt slightly. Then bury the nettles in good humus. Remove the upper layer of soil, put in the nettles and cover them with a layer of peat to protect the nettles, and then cover with the removed soil. After a full year, take the fermented nettles out and put them in an earthenware vessel, which, like the other preparations, is stored in a cool cellar covered with peat.

Preparation 505: Oak bark

Scrape off some bark from a sturdy oak without harming the tree. A good handful or two is sufficient. The bark must then be chopped. Fill the crumbly mass into the skull of a cow or sheep, closing the skull with a bone. At the end of September, at Michaelmas, bury it in a muddy, wet place. If there is no suitable location, it can be created. Fill a vat with mud and rotten plant remains and set the skull in there, cover with mud and add rainwater. In spring, around Easter, remove the skull and take out its contents, put them in an earthenware vessel, and store like the other preparations.

Preparation 506: Dandelion

Pick dandelion flowers in the early morning hours when they are just beginning to open. Then dry them. In fall, wrap them into the mesentery of a cow, pressing them tightly in. Bind and tie up this package with string. Like the other preparations, bury it in late September, at Michaelmas, in 20–30 cm (8–12 in) of good soil, and take it out at Easter. Store in an earthenware vessel like the other preparations.

Preparation 507: Valerian

Pick valerian flowers in the morning. Press out their juice and put it into small bottles, which are stored in a cool cellar. This juice will keep for a long time.

Preparation 508: Horsetail (equisetum)

The whole plant is picked and dried. Its tea is used to prevent fungal attacks. This tea has to be boiled for an hour in order to loosen the silica. The concentrated tea is diluted 1:10 with water.

Before fungi-endangered crops (potatoes and grains) are planted, spray about 60 litres of tea per hectare (6½ gallons/ acre) on the soil. If the crop is already infected with fungus, spray it immediately and repeat the sprayings.

Generally it is advisable to work with the Biodynamic Sowing and Planting Calendar (Maria Thun, published annually by Floris Books).

Some Personal Experiences

Newly seeded pastures

One morning I brought the cattle — thirty cows and a large bull — to a fresh pasture. I went to bring them in for milking in the late afternoon. From the distance I did not see a single cow on the pasture. I wondered whether they had broken loose, for there was only an electric fence. When I came closer, I saw that the whole herd, including the bull, were lying on their sides as though dead. Of course I was shocked. However, the animals were alive. I drove them up and with difficulty they staggered to the farm. Their manure came out like water. In the evening after milking I brought them to another pasture. The next morning they were fine.

What had happened? In the preceding year I had newly sown the grass on this pasture. I used my own seed, but did not have enough. So I quickly bought a ready prepared mixture of seeds for the remaining area. Before seeding, a small amount of compost was spread on the field followed by horn manure preparation. In the following spring again the horn manure was applied and later on horn silica was sprayed. The first grazing I described was just before the grass went into flower.

After this frightening experience I searched for its cause. I found the paper showing the composition of the grass seed

I had bought. The greater portion was 'tetraploid' seed. I asked a Polish seed grower who was staying with us what the 'tetraploid' breeding process involved. She replied that the grass seeds used for breeding were treated with radio-active cobalt, and 98% of these seeds would no longer germinate. The remaining 2% was used for growing the seed. This is how these tetraploid grasses came about. They are in effect genetically manipulated, genetically changed.

Grass primarily consists of protein and these unnatural proteins caused these disastrous digestive troubles. No further comment is necessary.

Radioactivity from Chernobyl

After the reactor accident in Chernobyl in 1986, with the help of all apprentices and students, we immediately treated the arable land with horn manure and horn silica. We left all the animals on the pastures despite official advice to keep them inside and to feed them with the previous year's feed.

Three weeks later two scientists came with various instruments to measure the radioactivity. All fields and pastures were tested, all animals and all our products (including the milk) were tested. The results showed no real increase of radioactivity beyond natural radiation which is always present. Measurements at our neighbour's farm showed a radiation increase of 13%.

The result on our farm was unbelievable for the scientists. One of them kept returning to our farm to find out how this could be explained. He observed the extraordinary effect of our spray preparations, and started to experiment with them himself, confirming their efficacy.

Dairy produce

As there were many apprentices and students wanting to work on our farm we had to find sufficient work for them. So we decided to process our milk ourselves. This led to great difficulties because the raw milk would not change in the right manner. Quark, cottage cheese, soft and hard cheeses did not turn out as they should. We then removed the milk vacuum pumps with the pipes and the automatic steam cleaning system. This brought better results in cheese making, yet not satisfying. We then followed the advice of Swiss farmers, who advocated to stop feeding silage. The results seemed like magic. We no longer had any difficulties at all in making quark, yoghurt, sour milk, soft and hard cheeses.

There was another unforeseen change: the flies in the cow barn, common today, vanished. With the change in feed there was no putrefaction in the first stomach of the cows. Flies need a putrid environment for breeding. As this no longer existed the flies dwindled.

The health of the cows also increased. The time between calves decreased to 330 days, and generally no assistance was needed when calving. The udders became markedly healthier and there was no longer excessive growth of the hooves. The most remarkable thing was the immediately improved state of health of the calves. They were raised effortlessly without signs of diarrhea, pneumonia, eczema, etc.

Barren land

The farm had an area that was totally barren, consisting of degenerated sandy soil. All attempts at raising crops on it failed. We then left it fallow. However, after a few years we tried again to till this field. This time a large amount of rotten manure was spread and we planted potatoes. The yield was fairly good and the fertility of this field increased continually, and finally even wheat was sown.

Birds and ants

When we acquired the farm in 1971, there were no ants and almost no birds on it. After landscaping it, creating pastures, meadows, ponds, hedges and fields, bird life greatly increased both in number and diversity. Recently an ornithologist found 53 kinds of birds in large numbers including rare threatened species.

After seven years sixteen ant heaps suddenly appeared on the farm. Only then did the farmland become highly fertile.

Rye

In the bakery on our farm we baked sour dough rye bread each year from 30 tonnes of rye flour (100 kg daily). That was the approximate annual yield of our rye. In one year we did not have enough rye, so we bought additional rye grain from another biodynamic farm. As we ground the flour, I became suspicious because during grinding the grain developed a very different smell compared to our grain. When we continued in processing, the dough developed such an obnoxious odour during fermentation that I wondered how nourishing this would be. To find the cause of this, I started to test other rye crops. The best results were from a farm working without animals. The reason was that silage feed for cattle created a dung that was unsuitable for field crops.

Milk yields

One winter when our herd was still quite young and some of the cows were dry, we delivered about 40 litres of milk to the dairy each day. Our neighbour delivered 200 litres a day. At the end of the month we both got our accounts from the dairy. My neighbour came to compare his bill because he knew that I did not use protein or other concentrated feed. The net sum I received was definitely higher than his. How could that be? He bought his concentrated feed, which was to increase the milk quantity of his cows, at the farmers' coop and they subtracted the price from his milk earnings. We both learned a lot from this experience.

Conclusion

To create a varied, versatile and flexible farm, it is important to get to know the complex and diverse ways of nature, and to work in harmony with the forces of nature and of the cosmos. A diverse farm environment can be created around the fields — woodlands, hedges, meadows, ponds, pastures, shrubs and orchards. The world of animals will enter these surroundings and find their right place and proper relationships to each other. Birds and insects are of special importance, and particularly earthworms who regulate the etheric forces in the soil. The harmonious interaction of the various forces of nature results in a wonderful balance, which enables the production of wholesome and healing nourishment for human beings. Not only is healing nourishment created, but the farm organism is itself a healing factor for the landscape and for the whole earth.

In order to produce healthy food, heal the earth and rebuild the soil and landscape, the present tendency, due to political and economic pressure, to create ever larger units and farms must be resisted. It is impossible to cultivate such huge units in harmony with nature and spirit. The worldwide destruction of soil is primarily caused by these giant farming units.

A true agriculture of the future must be based on small units — or larger units run by groups of people with the same ideals, perhaps including institutions for people with special needs.

What can we do now?

We must wake up and continually strive to search for truth. Our insights must be spread as widely as possible so that a large-scale fundamental change can be brought to our treatment of the earth. For earth and soil are the fundamental basis for our physical existence, and their health is essential for our survival. If we continue with one-sided materialistic thinking in our treatment of agricultural land, the soil will die and the extinction of humankind will only be a matter of time. Every person aware of this problem should help to bring about a change. How can this be done?

We should choose only biodynamic (or at least organic) foods, for they are imbued with vitality and cosmic forces due to the way they are produced. These forces promote health of body as well as of mind and spirit.

This in turn would guarantee the economic existence of biodynamic farms, and the growing demand for biodynamic products would encourage conventional farmers to convert. Biodynamic farms also contribute to the well-being and healing of the whole earth.

This continual giving and taking helps us out of our self-centred, egotistical mindset to an altruistic way of thinking which benefits the greater community. Every person really trying to act out of knowledge and wisdom betters the whole of humanity. This is something that every individual can do.

Agriculture must be free of governmental intervention, regulations and prescriptions, for the experienced and wise farmer is aware of the needs of his land in its varying conditions. Therefore he needs spiritual freedom. Nature is not a machine made by man to function as we want her to. She cannot be manipulated; she is not subject to static laws but to her own intrinsic spiritual, cosmic laws.

Our legal system must give greater equality. Equal rights are for all not just for the powerful. It must be possible for a farmer suing a chemical giant for damages to be taken seriously.

In our economic life we need more mutual, fraternal aid. If the soil is to be worked and cared for in the best way, there must be a sufficient number of people willing to do this. Perhaps some of the many young people who cannot find work or want to move away from the stress of working for large corporations might find their vocation in this care for the land. They might benefit from this emotionally and broaden their horizons. Some of these people may become enthusiastic about agriculture and healing the earth, and decide to spend their life doing this work. This would bring greater harmony and health to our social life.

A spiritual battle is currently occurring between destructive materialism and a practical idealism which works constructively to heal the earth, in harmony with nature and the cosmos.

References

1 Steiner, Agriculture, London 1974. Lecture of June 7, 1924, p. 23.
2 Steiner, as above, Lecture of June 10, 1924, p. 29.
3 Steiner, Spiritual Foundations for the Renewal of Agriculture, Kimberton 1993. Report of June 20, 1924, p. 3
4 Steiner, as above, Lecture of June 10, 1924, p. 40.
5 Thomas Meyer, Ein Leben für den Geist: Ehrenfried Pfeiffer, Basle 1999. p. 148.
6 Fritz Albert Popp, Biophotonen, das Licht unser Zellen, 1975.
7 Ehrenfried Pfeiffer, lecture of October 1, 1958 on nutrition.
8 André Voisin, Soil, Grass and Cancer, Austin, Texas 2000.
9 Hagalis Laboratory for crystal analysis und quality management, Überlingen, Germany.
10 Peter Tompkins and Christopher Bird, The Secret Life of Plants, Harper & Row, 1973, p. 154.
11 Peter Tompkins and Christopher Bird, Secrets of the Soil, Harper & Row, 1989, p. 135–38.
12 Welt am Sonntag, Aug 24, 1997 and March 18, 2001.

The Biodynamic Sowing and Planting Calendar 2017

Maria and Matthias Thun

This useful guide shows the optimum days for sowing, pruning and harvesting various plants and crops, as well as working with bees. It is presented in colour with clear symbols and explanations.

The calendar includes a pullout wallchart which can be pinned up in a barn, shed or greenhouse as a handy quick reference.